FIRE PERFORMANCE OF EXTERNAL THERMAL INSULATION FOR WALLS OF MULTISTOREY BUILDINGS

Third edition

Sarah Colwell and Tony Baker

bretrust

This work has been funded by BRE Trust. Any views expressed are not necessarily those of BRE Trust. While every effort is made to ensure the accuracy and quality of information and guidance when it is first published, BRE Trust can take no responsibility for the subsequent use of this information, nor for any errors or omissions it may contain.

The mission of BRE Trust is 'Through education and research to promote and support excellence and innovation in the built environment for the benefit of all'. Through its research programmes BRE Trust aims to achieve:

- a higher quality built environment
- built facilities that offer improved functionality and value for money
- a more efficient and sustainable construction sector, with
- a higher level of innovative practice.

A further aim of BRE Trust is to stimulate debate on challenges and opportunities in the built environment.

BRE Trust is a company limited by guarantee, registered in England and Wales (no. 3282856) and registered as a charity in England (no. 1092193) and in Scotland (no. SC039320).

Registered Office: Bucknalls Lane, Garston, Watford, Herts WD25 9XX

BRE Trust
Garston, Watford WD25 9XX
Tel: 01923 664743
Email: secretary@bretrust.co.uk
www.bretrust.org.uk

BRE Trust and BRE publications are available from
www.brebookshop.com
or
IHS BRE Press
Willoughby Road
Bracknell RG12 8FB
Tel: 01344 328038
Fax: 01344 328005
Email: brepress@ihs.com

BR 135

First published 1988
Second edition published 2003
Third edition published 2013
ISBN 978-1-84806-234-4

Printed on paper sourced from responsibly managed forests

CONTENTS

Cont'd . . .

EXECUTIVE SUMMARY

The risk of fire spread in multistorey buildings is an issue of concern, and recent fires have continued to highlight this. Since the external cladding system of the building offers one potential route for fire spread through a multistorey building there is a need for guidance to address these concerns. This latest edition of BR 135 – *Fire performance of external thermal insulation for walls of multistorey buildings* presents revised guidance that, while continuing to address the principles and design methodologies related to the fire-spread performance characteristics of non-loadbearing external cladding systems, also considers the changing drivers in this market, such as the recent increase in the new-build market for these types of system, and increasing thermal performance requirements. This new edition of BR 135 seeks to bring together the experience gained in this area by updating the guidance on external fire performance for the materials and technologies now used in the construction of external cladding systems.

The first edition of BR 135 was published in 1988 in response to the increasing use of thermal insulation as part of refurbishment programmes on existing multistorey residential tower blocks. The guidance presented detailed design solutions based on the range of products in the marketplace at the time. It is an interesting document to review when considering the current levels and types of thermal insulation system employed 25 years later, and being addressed by the third edition of this document.

The illustrations and scenarios presented in this current edition are based on typical examples of current practice, but, as has already been identified, this field is subject to rapidly changing designs and materials, and so this guidance focuses on the issues surrounding the topic, to enable designers and end users to understand better the parameters impacting on the fire-safe design and construction of external cladding systems.

The third edition also consolidates the fire performance classification systems for the full-scale fire tests in the BS 8414 series and found in Annex A of the second edition and Annex B, which was published as Digest 501, into a single document.

As part of the revision process for this document, a consultation exercise was undertaken with the key stakeholders to try to ensure that their experiences and issues in this rapidly changing market were addressed. The level of engagement and support from these stakeholders has been encouraging, and their input is reflected in the increasing range of systems identified and described in this third edition of the document, which we hope will continue to provide useful guidance in this field.

1 INTRODUCTION

The first edition of BR 135 – *Fire performance of external thermal insulation for walls of multistorey buildings* was published in 1988[1] in response to the increasing use of thermal insulation as part of refurbishment programmes on existing multistorey residential tower blocks. The guidance presented detailed design solutions based on the range of products in the marketplace at the time. It is an interesting document to review when considering the current levels and types of thermal insulation system employed 25 years later, and being addressed by the 3rd edition of this document.

At the time that the first edition was produced, there was no standard full-scale fire test available, and the test work behind the guidance was based on a single-faced, large-scale test facility similar to the test facility that now forms the basis of the BS 8414 test series[2,3], but without the wing return wall.

During this period a fire occurred in a refurbished block of residential flats in Liverpool. The Knowsley Heights fire in 1991 (Figure 1) suggested that a full-scale fire test method was necessary to fully understand the overall fire performance of the complete system as installed in these applications, using a representative fire scenario rather than relying solely on an elemental approach to try to control the overall fire performance of the system.

Additionally, the range of materials and potential design solutions available in the market was beginning to change, and was falling outside the range of guidance available in the first edition. As a result of this need to review the guidance and develop a full-scale test method, the then Department of the Environment worked with industry in a collaborative project to develop a full-scale test method, which was subsequently published by BRE in 1999 as Fire Note 9 – *Test method to assess the fire performance of external cladding systems*[4].

In June 1999, a fatal fire occurred in a multistorey residential housing block in Scotland (Figure 2). A resulting parliamentary inquiry was undertaken by the Environment Subcommittee of the Environment, Transport and Regional Affairs Committee to investigate the potential risk of fire spread in buildings by way of external cladding systems. As part of their recommendations[5], the subcommittee asked that the relevant guidance in

Figure 1: Knowsley Heights

Figure 2: Garnock Court, Irvine

this area be reviewed. This review process resulted in the publication of the second edition of BR 135 in 2003[6], and was accompanied by the publication of the full-scale fire test method from Fire Note 9[4] as BS 8414-1 *Fire performance of external cladding systems – Part 1: Test method for non-loadbearing external cladding systems applied to the face of the building*[2].

With the publication of BS 8414-1, a classification of fire performance was also required in order to interpret the test data: this classification system formed Annex A of the second edition of BR135[6]. Both BS 8414-1 and the associated classification method in Annex A of the second edition of BR 135 were developed to address systems applied to a masonry substrate.

Following the publication of the second edition of BR 135 in 2003, it became apparent that the technology being utilised for the construction of cladding and external wall systems was developing in such a way that the BS 8414-1 test method was no longer appropriate for assessing the fire performance of design details and systems that utilise framed wall structures. This resulted in the publication of BS 8414-2 – *Fire performance of external cladding systems – Part 2: Test method for non-loadbearing external cladding systems fixed to and supported by a structural steel frame* in 2005[3], and the subsequent publication of the classification system for this standard as Annex B to BR135, which was published as BRE Digest 501 – *Performance criteria and classification method for BS 8414-2* in 2007[7].

This third edition of BR 135 consolidates Annexes A and B into a single document, and supersedes BRE Digest 501. It also recognises the changing drivers in this market, with the recent increase in the new-build market for these types of system, and brings together the experience gained in this area by updating the guidance on the external fire performance for the materials and technologies now used in the construction of external cladding systems.

1.1 Scope

This guidance document addresses the principles and design methodologies related to the fire-spread performance characteristics of non-loadbearing external cladding systems. Although various potential design solutions have been identified and discussed in this document, robust design details are not presented, as it has been found that generic solutions are not available to this rapidly changing market, where new products and novel design solutions are frequently presented. The illustrations and scenarios presented in this document are based on typical examples of current practice, but, as has already been identified, this field is subject to rapidly changing designs and materials, and so this guidance focuses on the issues surrounding the topic to enable designers and end users to understand better the parameters impacting on the fire-safe design and construction of external cladding systems.

It should be noted that BS 8414-2 and Annex B of this guide relate specifically to external cladding systems applied to steel-frame constructions. At the time of writing this document, consideration has not been given to other construction systems such as concrete-frame or timber-frame constructions. However, the general principles given this report may still apply although suitable additional risk assessments and detail design reviews would be required.

1.2 Terms and definitions

Unless otherwise stated, the terminology used in this document refers back to either the definitions presented in the BS 6100 series of standards, *Building and civil engineering – Vocabulary* or EN ISO 13943:2000, *Fire safety – Vocabulary*[8].

2 LEGISLATION

2.1 Building Regulations

The Building Regulations in the UK[9-11] set out the requirements that must be met in relation to design and building work in the construction of domestic, commercial and industrial buildings. The regulatory systems in Scotland and Northern Ireland differ from those in England and Wales, but the underlying requirements are similar.

The Regulations and supporting guidance and standards set out acceptance criteria for a wide range of interrelated technical provisions. Care should be taken at the design stage, as the needs of one provision may conflict with the needs of another, and designers must be able to satisfy each provision without contravening another.

These potentially conflicting requirements are highlighted in the area of innovative materials and designs, which are being driven by the need to construct more energy-efficient and sustainable buildings. In order to meet these design challenges, the range of new and innovative materials and designs of systems being offered as potential solutions has also increased the volumes of potentially combustible materials being used in external cladding applications. In addition, these external wall cladding systems, while still widely used for refurbishment applications on masonry structures, are increasingly being applied to new-build scenarios utilising lightweight framing solutions for the wall system.

External cladding systems will generally need to address requirements that include:

- resistance to moisture/condensation
- wind loading
- ventilation
- thermal performance – conservation of fuel and power
- sustainability and durability
- fire performance
  - fire resistance (Part B3 *Internal fire spread (structure)* of Approved Document B, Volumes 1 and 2, in England and Wales; Section 2 of *Domestic and non-domestic technical handbooks* in Scotland; Section 3 of *Technical booklet E* in Northern Ireland)
  - external fire spread (Part B4 *External fire spread* of Approved Document B, Volumes 1 and 2, in England and Wales; *Section 2 of Domestic and non-domestic technical handbooks*; Section 4 of *Technical booklet E* in Northern Ireland).

Issues relating to the fire-resistance performance of external cladding systems, eg in relation to boundary conditions and space separation, are discussed in the BRE guide BR 187 – *External fire spread: building separation and boundary distances*[12], and are not addressed in this guide. Provisions for fire resistance can be found in the relevant sections of the documents referred to above.

This guide provides a basis for evaluating the fire performance of external cladding systems. It does not specify where this performance standard should be adopted; this is a matter for regulators and specifiers. However, the performance standard set out could be adopted where the implications of rapid fire spread by way of the external cladding system are considered to be unacceptable, such as tall buildings (with a storey height 18 m or more above ground level) that may be out of the reach of conventional firefighting techniques, and where external fire spread may present an unacceptable risk to the building's occupants.

2.2 Property protection

The Building Regulations address the design and fire performance requirements for external cladding systems in order to ensure the life safety of those associated with the building, but they do not address the property protection, business interruption and social consequences of a major property loss arising from a fire involving the external cladding system.

Fire performance standards such as the Loss Prevention Standards LPS 1581 – *Requirements and tests for LPCB approval of non-loadbearing external cladding systems applied to the masonry face of a building*[13] and LPS 1582 – *Requirements and tests for LPCB approval of non-loadbearing external cladding systems fixed to and supported by a structural steel frame*[14] are designed to provide users with third-party-approved external cladding systems for the purposes of property protection, and therefore they use different performance criteria, which go beyond the criteria set out in Annexes A and B of this document.

3 MECHANISMS OF FIRE SPREAD

The key stages associated with fire spread on the outside of a building envelope are:

- initiation of the fire event
- fire breakout
- interaction with external envelope
- fire re-entry
- fire service intervention.

These stages are discussed below, and are illustrated schematically in Figure 3.

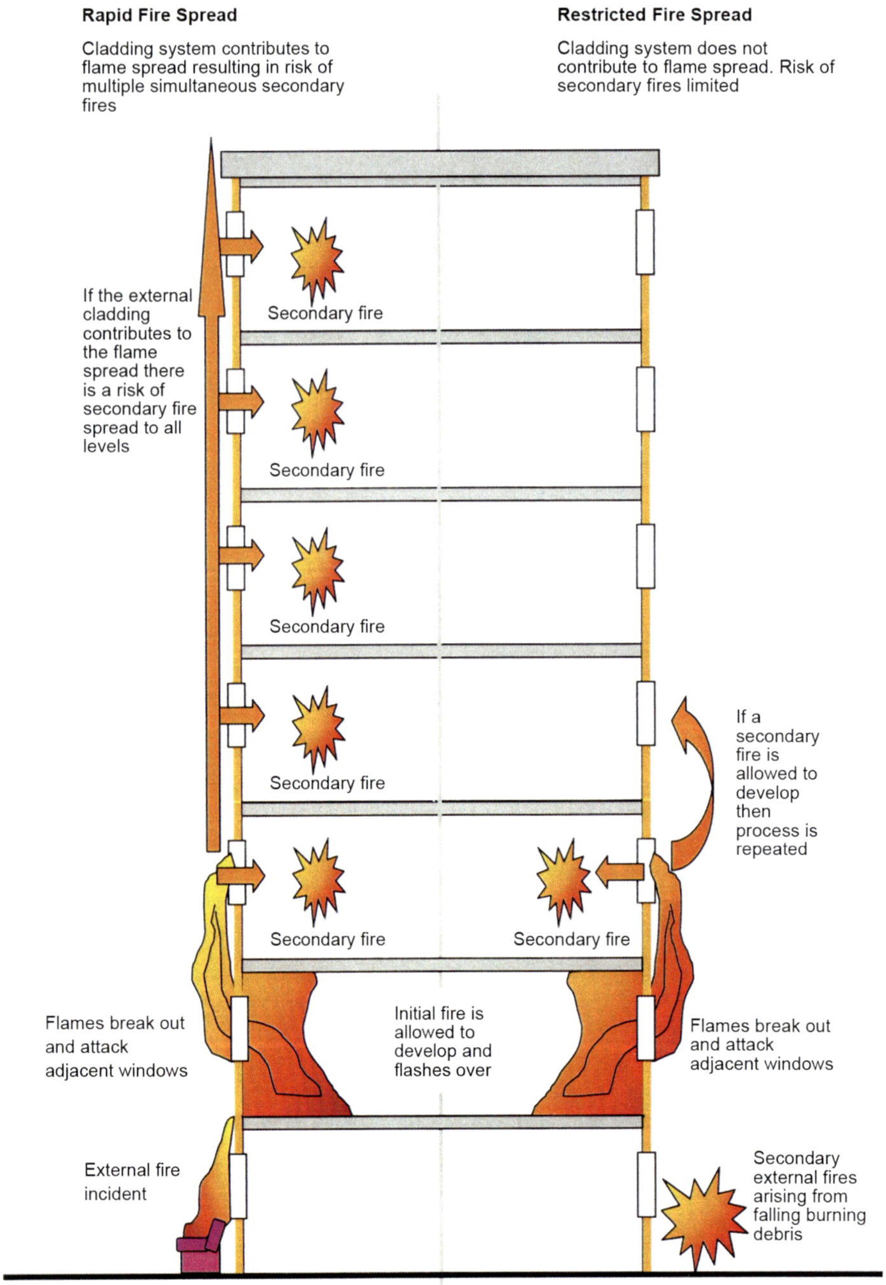

Figure 3: Mechanisms for external fire spread by way of the external cladding system

3.1 Initiation of the fire event

This type of fire event can be initiated from a fire occurring inside the building, or by an external fire in close proximity to the building envelope, such as fires involving general waste, or resulting from malicious firesetting.

3.2 Fire breakout

Following the initiation of a fire inside the building, if no intervention occurs, the fire may develop to flashover and break out from the room of origin through a window opening or doorway (Figure 4). Flames breaking out of a building from a post-flashover fire will typically extend 2 m above the top of the opening prior to any involvement of the external face, and this is therefore independent of the material used to construct the outer face of the building envelope (eg Figures 4 and 5).

3.3 Interaction with the external envelope

It is at this stage of the fire scenario that the fire performance of the complete external cladding system, including any fire barriers, is critically important. Once flames begin to impinge upon the external fabric of the building, from either an *internal* or an *external source*, there is the potential for the external cladding system to become involved, and to contribute to the external fire spread up the building by the following routes.

3.3.1 Surface propagation

The reaction to fire characteristics of the materials used within the external cladding system will influence the rate of fire spread up the building envelope by way of the surface of the external cladding system.

3.3.2 Cavities

Cavities may be incorporated within an external cladding system, or may be formed by the delamination or differential movement of the system in a fire. If flames become confined or restricted by entering cavities within the external cladding system, they will become elongated as they seek oxygen and fuel to support the combustion process. This process can lead to flame extension of five to ten times that of the original flame lengths, regardless of the materials used to line the cavities. This may enable fire to spread rapidly, unseen, through the external cladding system, if appropriate fire barriers have not been provided (Figure 6).

Figure 5: Fire damage

(a)

(b)

Figure 4: (a) Fire breakout from a post-flashover room on a masonry facade, (b) result of the flashover

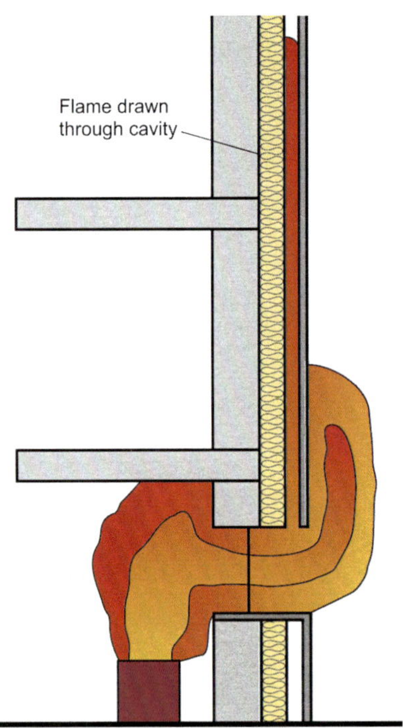

Figure 6: Fire spread through cavities

3.4 Fire re-entry

Window openings or other unprotected areas within the flame envelope provide a potential route for fire spread back into the building. This creates the potential for fire to bypass any compartment floors that may be present, leading to a secondary fire on the floor above. If secondary fires are allowed to develop without intervention before flashover occurs, then flames may break out again, thus extending the flame envelope and threatening other openings further up the building, irrespective of the materials used on the building envelope.

3.5 Fire service intervention

Where the external cladding system is not contributing significantly to the spread of fire from one storey to the next, then intervention by the emergency services should prevent continued fire propagation by way of the building envelope. However, where the external cladding system is contributing to the fire propagation rate, the potential exists for the fire to affect multiple storeys simultaneously, thus making firefighting more difficult.

4 CLADDING SYSTEMS: APPLICATION AND TYPES

External cladding systems fall into two distinct types of application:

- refurbishment of existing multistorey housing stock, or redevelopment of commercial or residential projects
- new build.

Each market has similar challenges; they both need to deliver enhanced thermal performance and sustainability.

The key difference between the two applications is not in the external finishes used, but in the substrate to which the systems are installed. The refurbishment market is typically based around systems that are installed on existing masonry-based substrates, such as blockwork infill panelled structures. The new-build market may also install systems to masonry substrates, but this sector is increasingly working with lightweight frame systems (LFS) based on either steel or concrete primary frames.

4.1 Materials for external finishes

The range of materials available for use as external finishes is large: from finishes such as stone, terracotta, concrete, timber or metal, which are typically used as part of rainscreen systems, through to insulated render systems that utilise a range of insulation products and finishes. As the technology has developed in this area, hybrid systems are increasingly being offered to the market that use a combination of materials, such as thin stone veneers resin-bonded to insulated carrier substrates, polymer-resin-bonded boards, and pre-insulated panels. Often several of these finishes will be applied in combination in a single project.

4.2 Glazing

Glazing systems can be found as a component in most facades, and are present in a variety of forms, such as infill panels, windows, and suspended glazing. Depending upon their end-use application, the systems may be formed as single- or double-glazed units. Where required, these systems can also be designed to meet the necessary fire-resistance parameters for their application.

4.3 Curtain walling

Curtain walling systems are not addressed in this guidance document; details of the fire testing of curtain walling systems can be found in the relevant CEN harmonised product standard for these systems, *BS EN 13830:2003 Curtain walling – Product Standard*[15].

4.4 Insulating materials

The range of insulation materials found in the external wall build-up can vary (see Table 1). For general refurbishment or masonry-based systems a single product is typically applied to the supporting structure, using a combination of mechanical and/or adhesive-based systems. On uneven or lightweight frame systems that require a cavity between the insulation layer and the sheathing boards, railing systems are typically applied; these will also require the selection of appropriate fire barriers to close the cavity in the event of a fire.

For lightweight frame systems, a combination of insulation products can also be found where one product is applied to the external face of the sheathing board and another is used as infill in the lightweight wall frame to provide both thermal and acoustic performance. As part of this type of design solution, vapour barriers and breather membranes will also be required to ensure the effective thermal operation of the wall system.

The fire performance of minor items such as breather membranes, vapour barriers and gaskets also needs to be considered; in isolation they may not pose a fire risk, but when used over large surface areas in multistorey dwellings they have the potential to contribute to fire spread on or through the system by offering a route to bypass fire prevention measures such as fire breaks and cavity closures.

Table 1: **Insulation products**

Generic insulation description	Examples	Applications
Non-combustible materials and materials of limited combustibility (as defined in Tables A6 and A7 of Approved Document B[9])	Generally mineral-fibre-based products such as stone fibre and glass wool, typically formed using a resin binder.	Produced in batts and rolls of various sizes. The thickness and density of these products can vary widely, depending on the thermal performance specification.
Thermoset products	Polyurethane foam (PUR), polyisocyanurate foam (PIR) and phenolic foams are part of this group, used to provide insulation for external cladding systems.	These products are provided in sheet form at various sizes and thicknesses to meet thermal performance requirements. They are often faced with materials such as glass fibre or aluminium foil.
Thermoplastic products	Expanded polystyrene (EPS) is the most widely used product in this group, which also includes extruded polystyrene (XPS). It can be supplied in both fire-retarded and non-fire-retarded forms.	The material is generally supplied in sheet form at various sizes and thicknesses to meet thermal performance requirements.
Natural fibres	Examples such as wood fibre, cork, sheep wool, cellulose and hemp are becoming increasingly widespread. The products are generally soaked, heated and compressed to produce the board product. In some cases binders are also used to provide the required performance characteristics.	The material is generally supplied in sheet form at various sizes and thicknesses to meet thermal performance requirements. These products may also be available as in situ fill products that are 'blown' into voids on site.
Recycled materials	A wide range of recycled materials, such as recycled paper and newsprint, shredded rubber and combinations of other materials, are available as insulation products, which may be treated or used with binders to achieve the required application and performance characteristics.	The form that these recycled products takes can vary from compressed boards to blown infills.

5 EXTERNAL FINISH CONSTRUCTION TYPES

Two system types have been identified for discussion here:

- non-ventilated systems
- ventilated cavity systems.

5.1 Non-ventilated applied finishes

These systems are typically used for refurbishment on masonry-based substrates where a continuous background structure, such as an external wall is being upgraded. The system is typically composed of two elements (Figure 7):

- The insulating material is typically fixed to the masonry background structure, to provide the necessary level of thermal performance.
- The external surface finish generally provides the weather protection to the insulating layer.

There are two distinct types of non-ventilated applied finish:

- The first type generally involves direct attachment of the insulation to a masonry background, or to a sheathing board/LFS background. The insulation may be adhesively bonded, adhesively bonded and mechanically fixed, or mechanically fixed either on rails or tracks or using proprietary insulation fasteners appropriate to the background type. The external finish is applied directly onto the external face of the insulation.
- The second type includes a drained, non-ventilated cavity behind the insulation. The insulation is typically fixed using rails or tracks, with packing provided between the rails and the background structure to form a cavity. Alternatively, the insulation may be attached to a sheathing board (as above) that itself is attached to the background walls on vertical rails or battens, thus creating a cavity that is drained, but not ventilated, behind the sheathing board. The external finish is applied directly onto the external face of the insulation.

5.2 Ventilated applied finishes

These systems typically consist of an inner structural leaf that is insulated on its outer face (Figure 8), and are finished with an external surface membrane or cladding assembly. Generally, the insulation layer is fixed to either the inner structural leaf or the cladding panel, together with an appropriate breather membrane. There are various different systems that fall within this description. They include rainscreen cladding systems and drained and ventilated cavity systems. They also include lightweight frame systems that need to maintain a cavity between the sheathing boards and the rear face of the insulation. These cavities are typically drained, back-ventilated or pressure-equalised systems.

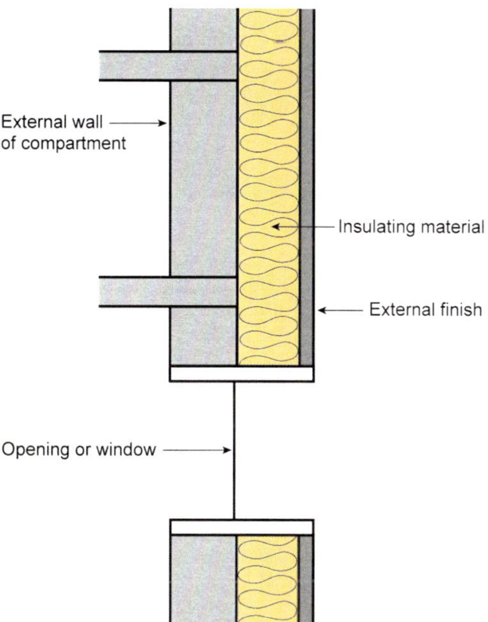

Figure 7: Non-ventilated system

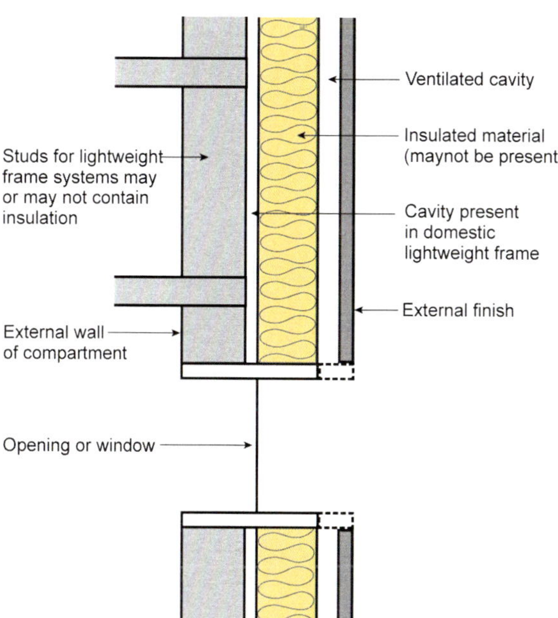

Figure 8: Ventilated cavity system

6 FIRE PERFORMANCE DESIGN PRINCIPLES FOR EXTERNAL CLADDING SYSTEMS

6.1 General

Generic fire performance design principles are presented in this guide for information. Robust design details are not provided, as experience from the earlier editions shows that this is a rapidly changing market, and that the details cannot keep pace with changing market needs. Also, experience from full-scale fire test programmes has shown that the wide range of applications over which external cladding systems are used do not lend themselves to robust design detail solutions.

In order to assist specifiers and designers in addressing some of the fire performance issues, a set of design principles has been developed, based on full-scale research programmes. These principles represent key elements that have been identified, but they are not exhaustive; other elements and principles may be worthy of consideration. General fire performance guidance for external cladding systems is also given in Approved Document B[9] and Section 2 of *Technical Handbooks*[10] or Technical Booklet E[11] as appropriate. Currently, innovative designs and variations in material selection and design can be fully assessed only by full-scale testing.

The following general points should be considered when designing and specifying external cladding systems.

- The installation and fixing methods employed should be sufficiently robust to withstand the potential thermal exposure and fire-spread characteristics associated with an external fire exposure without exhibiting significant fire spread or system collapse, such as:
  - loss of strength as the system is heated
  - forces generated by retained thermal expansion of fixings and components
  - movements and distortions arising from thermal expansion if unrestrained.

  These details are of particular concern when LFS are used. Also, the robustness of the subframes and their interconnection to the primary structure is of critical concern in relation to the overall fire performance of the external cladding system.
- The fire performance characteristics of the external finish should be considered with the substrate to which the finish is applied, to ensure that the system does not provide a route for significant fire propagation.
- The system should not delaminate or spall to provide a potential entry route for fire to access unprotected cavities or combustible materials.

6.2 Supporting substrates and frameworks

As outlined earlier, the two primary systems considered in this guide are:

- masonry-backed systems – see BS 8414-1[2] and Annex A of this document
- lightweight frame systems – see BS 8414-2[3] and Annex B of this document.

In general, masonry walls offer a degree of thermal stability and fire resistance during a fire, and therefore it is only the external cladding system fitted to this substrate that is considered in this guidance.

The frame-based wall systems (from internal room face to external weather face) offer an additional level of complexity, in that the stability of the overall system is also based on the wall frame's ability to remain in place during a fire. Also, any deformation or degradation of the framed wall may have a detrimental effect on the fire performance of the external cladding system. Therefore, when determining the fire performance of an external cladding system secured to the face of a framed wall system, the performance of the complete wall system from the internal room face to the outer surface of the weatherproofing system must be considered.

These systems are typically constructed on either a steel or concrete primary loadbearing frame, with either lightweight steel infill stud frames that are fitted between floor slabs, or as a continuous frame oversailing the ends of the floor slabs and held back to the main steel or concrete frame at the floor slab edge details. In either case the stud frames are typically lined internally with plasterboard sheets to provide fire and acoustic performance; the cavity between the studs can also be filled with insulation to improve the fire, thermal or acoustic performance, and this will vary on each design application. The outer face of the wall frame is typically fitted with a sheathing board on which the external insulation and finish are installed.

Where these stud walls are used for domestic residential applications, some designs require the provision of a cavity, typically 15–20 mm wide, between the sheathing board and the insulation layer making up the external cladding system. When designing the fire protection for these systems, this additional cavity, and any variation in the width of the cavity resulting from building tolerances, must also be taken into account.

6.3 System-specific details: rendered systems

As the requirements to improve the thermal performance of buildings increase, so the volume of insulation material used within buildings also increases. Where rendered insulation systems are used on the outer face of the building envelope, the typical performance of these systems in fire can be summarised as follows.

The area of external cladding within the flame envelope will typically begin to become friable as any volatile components are driven off. As the system continues to be heated it will expand, and the external render finish may delaminate from the insulation and primary structure of the building envelope as it is heated by the fire source (Figure 5).

As the external render finish begins to break down and move under the applied heat it will tend to develop cracks and fissures in its surface, allowing the fire to penetrate the external finish and attack the insulation layer beneath. If the external finish and any supporting mesh are not adequately restrained as the system expands, the fixings may become detached from the insulation layer and the primary substrate, and the system will begin to delaminate and fall away. As the system exposes increasing quantities of insulation material, this may lead to increased fire spread through the system, together with the production of falling debris and the potential for system collapse. In LFS, if the external insulation layer is either consumed or falls away, the sheathing boards or subframe can then become exposed to the fire load. The potential then exists for the fire to break through the system and continue to spread.

6.3.1 Fixing details

To prevent rapid fire spread and system collapse, all external cladding systems should be installed with suitable through-fixing methods to ensure that the system will not suffer disproportionate collapse during a fire. The systems are typically fixed to the supporting structure using a combination of mechanical fixings (both plastic and steel) and adhesives.

Adhesive-based systems

The increasing quantities and therefore weight of insulation material used within render systems may preclude the future use of solely adhesive-based systems. Dependent upon the fixing details, the systems can be unstable in fire if they have no mechanical restraint from the external finish to the wall.

If the insulating material is consumed or degraded during a fire, then the adhesion within the system will be lost, allowing the external finish to move independently of the wall, and giving rise to potential system delamination or collapse. An adhesive-based fixing method can be supplemented with mechanical fixing to provide increased system stability.

Mechanical fixings

There are several mechanical fixing methods available. Not only must these systems meet general service requirements, such as resistance to thermal bridging and corrosivity, they must also have the mechanical strength to support the external insulation and cladding systems under normal operating conditions. They should also be capable of holding the systems in place in the event of fire to prevent excessive fire spread and system collapse. The fixing systems include complete steel or plastics systems, together with combination systems that use plastic carriers with steel fixing pins.

The guidance given in BRE Defect Action Sheet DAS 132[16] should be noted:

'Use no fewer than one stainless steel fixing (in addition to those of plastics) per square metre of insulation. The fixings should be sized and fitted to resist the increased duty that may be required under fire conditions.'

Although the insulation may be adequately held to the substrate, the stability of the finish coats should also be considered in order to avoid excessive system delamination, which may generate voids into which the fire may spread. For this purpose, the use of mechanical fixings to attach the base coats to the substrate or through-fixing details at fire barriers should be considered.

For LFS, where the insulation is fixed to the substrate with a rail system to provide a drainage cavity, it is important that the rail system is adequately fixed so that the insulation is fully supported in the case of a fire, and does not allow the product or render to collapse. The way in which the systems are fixed and the load is transferred to the LFS or supporting framework should be considered to ensure that there is no disproportionate collapse or movement of the system during a fire when the system moves or distorts as it is heated.

6.3.2 Fire barriers

If fire enters a void in the system, whether that void is created by a fire or is part of an existing design, and the insulation is exposed to the fire source, any combustible material present may become involved, and there is potential for the fire to propagate throughout the system if adequate fire barriers are not installed. Since a cavity is likely to be present behind the insulation boards in LFS, it is important that this potential is recognised, and adequate fire-stopping is provided, using fire barriers or fire-stopping details to maintain the system's stability in the case of fire.

6.3.3 Insulation systems

The relative fire performance of the generic insulation materials presented in Table 1 can be summarised as follows.

Non-combustible materials and materials of limited combustibility

The definitions of non-combustible materials and materials of limited combustibility for the purposes of this guide are those given in Tables A6 and A7 of Approved Document B[9].

All material within the fire envelope will be damaged during the course of a fire. Stone mineral wool based products tend to lose some integrity, but the material typically remains intact. Although glass mineral wool material does not exhibit fire propagation, if it has been directly exposed to the fire source it may become degraded, and in some cases may melt away. An example is shown in Figure 9.

Thermoset products

Unless such materials become directly exposed to the fire source, following significant delamination and cracking of the external render finish, they will typically char in the vicinity of the fire source. If the insulation is directly exposed to the fire source, and adequate fire barriers are not installed, fire spread may arise on the surface of the exposed insulation, allowing additional cracking and delamination of the external render coat to occur, and so providing a route for continued fire spread through the system away from the initial fire source (Figure 10).

Thermoplastic products

Thermoplastic products such as expanded polystyrene (EPS) will typically soften and melt in the early stages of a fire, generating a void behind the external render finish coat. If inadequate fixings have been used, without the support of the insulating material the finish coats will sag and crack, producing a direct entry route for the fire to the insulation material. Once the material ignites, rapid fire spread can occur if suitable fire barriers and fixing details are not provided. The relatively low softening and melting points of EPS mean that damage can occur to the insulation layer well away from the seat of the fire.

Figure 11(a) shows a fire test of an EPS system without fire barriers and Figure 11(b) shows a test with fire barriers fitted.

Natural fibres

This group of products covers a wide range of materials, some of which will have low surface spread of flame characteristics and others relatively high. Typically, though, these products will not tend to burn away rapidly to leave voids behind the render systems during the initial fire event or in the immediate aftermath. They tend to exhibit a degree of damage in the fire plume area, with some loss of system integrity. A key issue with these products is the potential for the insulation material within the fire plume to exhibit localised, glowing hot spots. These areas of localised combustion can continue to smoulder and burn for many hours after exposure to the fire load, and can be difficult to locate. This leads to a potential risk of fire propagating unseen behind the render face. This is not necessarily a life safety issue for masonry-backed systems, but there is a potential for unseen fire spread and re-entry into the interior of a building in LFS, and this should be considered as part of any risk assessment associated with the use of these products.

6.3.4 Design principles for fire barriers: render systems

Various full-scale experimental studies have shown that for rendered systems to meet the performance criteria set

Figure 9: System with insulation comprising a material of limited combustibility, after test

Figure 10: Thermosetting core without adequate fixings or fire breaks

(a)

(b)

Figure 11: EPS system after test (a) without fire barriers, (b) with fire barriers

out in this guide, any fire travelling through the system should be contained to the floor level immediately above the fire origin. To achieve this, the installation of fire barriers at each floor level above the first floor level (ie starting with the second storey) should be considered, or at alternative spacing determined by full-scale testing to the BS 8414 series[2,3].

Design details such as those provided in Detect Action Sheet DAS 132[16] have been available for some time; they offer one potential solution for masonry-backed systems or non-domestic frame systems that do not require the provision of a drained or ventilated cavity. The key design elements for these fire barriers as given in DAS 132 are as follows:

- The fire barrier should be at least 100 mm high, and should form a continuous band through the insulation layer at each floor level. Any abutting of material should ensure that no cavity exists for fire to track or pass through.
- The fire barrier should be formed from non-combustible material as defined in Table A6 of Approved Document B[9], and should be bonded and tied back to the wall and the external render finish to ensure that no fire path can be created between the non-combustible material and the primary substrate, or between the non-combustible material and the external render finish.
- Through-fixing of the render base coat to the primary substrate with all-steel fixings should also be considered, to ensure that no movement of the external render finish away from the fire barrier is possible. It is important that there is no potential for fire between the external render coat and the fire barrier.

With the increasing use of external cladding systems and the experience of their performance in fire, several alternative proprietary fire barrier systems are now available. Their suitability for a given design scenario requires detailed consideration by both specifiers and suppliers. In some LFS a cavity (typically 15–20 mm) is required between the sheathing board and the insulation layer. In order to maintain this cavity during normal activities, an intumescent-based system that can close the cavity in the event of a fire is typically employed. These narrow cavities are also seen in rail-based insulation systems used to address situations such as uneven substrate surfaces; again, intumescent-based systems are frequently employed.

The use of fire barriers should also be considered for application in vulnerable areas such as window openings and doorways, and around penetrations in the system.

Vertical fire barriers may also be required to prevent lateral fire spread when it is necessary to maintain compartmentation in the structure, or to prevent excessive fire spread for property protection applications.

Although intermediate-scale testing can assist in assessing and selecting fire barriers, the overall effectiveness of the barrier design can be fully assessed only as part of a system test at large scale.

6.4 Ventilated cavity systems

Figure 8 shows typical design details for ventilated cavity systems. The external wall panel supports are generally constructed from timber battens or metal railing systems. Aluminium and polymeric railing systems are predominantly used because of their relatively light weight and ease of maintenance. The walls are typically fitted with insulating material laid between the support railings, and the external panels are fitted to the railing system, leaving a ventilation cavity between the panels and the insulation.

If the fire is able to enter the cavity, it may propagate unseen through the system if adequate fire barriers are not employed. This may result in significant risk of system collapse, or in the fire breaking out at significant distances from its origin.

In LFS with a secondary cavity between the sheathing board and the insulation, the design of the fire protection can be challenging, as it needs to meet conflicting requirements: to maintain air flow through the system for normal duty, but to prevent flame spread in the event of a fire.

In order to counter the possibility of rapid fire spread and potential system collapse, the design and selection of materials used to construct these systems should address these issues, including the provision of fire barriers.

6.4.1 Performance of materials in fire

Insulation

As it can be difficult to prevent fire entering the cavity and spreading in these systems, the selection of the insulation materials used and the design of the fire barriers to close these cavities are particularly important. In LFS the location and types of insulation used will also depend on the thermal, acoustic and fire performance design of the system. Once the fire enters the cavity, the surface of the insulation materials is exposed to the fire source. The fire is contained within a narrow cavity, which will encourage elongation of the flame front, and so increase the potential for propagation of the flames through the cavity. It is therefore essential to provide suitable fire barriers to protect these systems.

Railing system

The railing system is used to carry the external panels and finishes for the cladding system, such as polymeric panels, ceramics and metal alloys. The temperatures within the fire envelope (Figure 3) may achieve local temperatures in excess of 600 °C: therefore, regardless of the external panel construction, if fire enters the cavity and comes into contact with the railing system, the system will begin to lose its local strength and integrity as it is heated. Under prolonged fire exposure, the structural integrity of the railing system may be affected, leading to system collapse regardless of the characteristics of the system's reaction to fire.

If timber or other combustible railing systems become involved in the fire, they may allow fire to propagate through the system if adequate fire barriers are not installed. It is technically challenging to provide suitable fire barriers for these types of system, as the potential propagation route through the system must be broken without comprising the strength and integrity of the railing system. Again, regardless of the fire performance characteristics of the external panels, if the railing system fails, this may lead to detachment of the external panels or collapse of the system.

External panels and finishes

The materials used for external panels used can vary from non-combustible through to combustible.

Non-combustible materials and materials of limited combustibility (as defined in Tables A6 and A7 of Approved Document B[9]) typically range from cementitious-based products and coated metal panels through to natural products such as stone veneers. Cementitious and stone-based products tend to spall and crack within the fire envelope, providing access for the fire to the cavity. There is a potential risk of injury to people or damage to property if spalling material is expelled from the system during a fire. The panels may also generate large pieces of falling debris if the integrity of the fixings to the railing system is lost during the fire. Metal panels may fall from the system if the strength of the fixings is affected by the local fire source. They may also melt, generating molten metal debris if exposed directly to the sustained flame envelope or other combustible materials in the system.

Combustible panels are typically based on vinyl or glass-reinforced plastic, although various new products are being developed in this area, some of which also contain insulation materials. These products generally have good surface spread of flame characteristics to prevent rapid fire spread across the surface of the system, but once the panels become involved in the fire, they have the potential to generate falling debris, add to the overall fire load, and provide a route for fire to propagate up the outside of the building.

6.4.2 Fire barriers: ventilated cavities

Fire barriers installed in ventilated cavity systems are intended to prevent fire propagation through the cavities and any combustible materials used within the system, while maintaining an air flow through the system that allows the cavity to operate effectively during normal circumstances. For LFS more than one ventilated cavity may be present in the system, which will add to the complexity of any design solution.

Various fire barrier designs have been proposed, including intumescent grill systems and through-fixed steel plates, but the key elements for producing an effective fire barrier for ventilated-cavity systems have been found to be:

- the details of the fixing of the fire barrier to the system substrate
- that the fitting for the fire barrier is independent of the sheeting rails
- that the fire barrier, when operating, closes across the full depth of the cavity and in some cases protrudes from the front face, to allow for movement of the panels during test
- that the fire barrier, when operating, closes against a non-combustible structure within the system such as a mineral fibre fire break.

The nature of the fire barriers required to prevent fire spread has been found to depend, in the main, on the nature of the cladding system itself. Limited experience has shown that effective fire barriers can be designed and installed for these systems. The fire barriers required the vertical sheeting rails to be cut, and therefore interrupted, at regular intervals. Certain barrier systems were found to be adequate for some sheeting materials but not for others. Fire barrier systems therefore need to be considered in the context of the complete system for each specific design, as currently there are no generic solutions that are suitable for all applications.

In practice it has been found that small-scale tests do not fully characterise the fire hazard associated with full-scale cladding systems. The only effective way to assess the fire performance of the fire barriers for this type of relatively complex system is to test the complete system at large scale.

The use of fire protection solely around the windows was generally found to be inadequate in preventing fire spread.

REFERENCES

1 Rogowski B F W, Ramaprasad R and Southern J R. Fire performance of external thermal insulation for walls of multi-storey buildings. BRE BR 135. Bracknell, IHS BRE Press, 1988.

2 BSI. Fire performance of external cladding systems. Part 1: Test method for non-loadbearing external cladding systems applied to the face the building. BS 8414-1:2002. London, BSI, 2002.

3 BSI. Fire performance of external cladding systems. Part 2: Test method for non-loadbearing external cladding systems fixed to and supported by a structural steel frame. BS 8414-2:2005. London, BSI, 2005.

4 Morris W, Colwell S, Smith D J and Andrews A. Test method to assess the fire performance of external cladding systems. Fire Note 9. Bracknell, IHS BRE Press, 1999.

5 House of Commons Environment, Transport and Regional Affairs Committee. Potential risk of fire spread in buildings via external cladding systems. 14 December 1999 (HC109).

6 Colwell S and Martin B. Fire performance of external thermal insulation for walls of multi-storey buildings. BRE BR 135. Bracknell, IHS BRE Press, 2003, 2nd edn.

7 Colwell S. BR135: Annex B Performance criteria and classification method for BS 8414-2. BRE DG 501. Bracknell, IHS BRE Press, 2007.

8 BSI. Fire safety. Vocabulary. BS EN ISO 13943:2010. London, BSI, 2010.

9 Department for Communities and Local Government (DCLG). The Building Regulations (England & Wales) 2010. Approved Document B: Fire safety, Volumes 1 and 2, 2006 edition. London, TSO, 2006. Available from www.planningportal.gov.uk and www.thenbs.com/buildingregs.

10 Scottish Government. Technical Standards for compliance with the Building (Scotland) Regulations 2004: Technical Handbooks, Domestic and Non-domestic, 2010: Section 2 Fire. Edinburgh, Scottish Government. Available from www.scotland.gov.uk.

11 Department of Finance and Personnel Northern Ireland. Building Regulations (Northern Ireland) 2012. Guidance: Technical Booklet E: Fire safety, 2012 edition. London, The Stationery Office. Available from www.dfpni.gov.uk.

12 Read R E H. External fire spread: building separation and boundary distances. BRE BR 187. Bracknell, IHS BRE Press, 1991. *Note: A 2nd edition is due for publication in 2013.*

13 Loss Prevention Certification Board. Requirements and tests for LPCB approval of non-loadbearing external cladding systems applied to the masonry face of a building. LPS 1581. Watford, BRE Global, 2007.

14 Loss Prevention Certification Board. Requirements and tests for LPCB approval of non-loadbearing external cladding systems fixed to and supported by a structural steel frame. LPS 1582. Watford, BRE Global, 2007.

15 BSI. Curtain Walling – Product standard. BS EN 13830:2003. London, BSI, 2003.

16 BRE. External walls: external combustible plastics insulation: fixings. BRE DAS 132. Bracknell, IHS BRE Press, 1989.

ANNEX A: PERFORMANCE CRITERIA AND CLASSIFICATION METHOD FOR BS 8414-1

This annex provides a classification system for the test methodology outlined in BS 8414-1 *Fire performance of external cladding systems – Part 1: Test method for non-loadbearing external cladding systems applied to the face of the building*[A1].

A1 TEST METHOD

Figure A1: Example of a typical test facility

Definitions

Level 1
A height 2 m above the top of the combustion chamber opening (Figure A3).

Level 2
A height 5 m above the top of the combustion chamber opening (Figure A3).

Start temperature, T_s
The mean temperature of the thermocouples at level 1 (Figure A3) during the 5 min before ignition.

Start time, t_s
The time when the temperature recorded by any external thermocouple at level 1 equals or exceeds a 200 °C temperature rise above Ts, and remains above this value for at least 30 s (Figures A3 and A5).

System
The complete cladding assembly, including any sheeting rails, cavities, fire barriers and weathering membranes or coatings.

A1.1 Principle of test

The test facility has been designed to allow the external fire performance of both applied and supported non-loadbearing external cladding systems to be determined (Figure A1).

The test facility allows external cladding systems to be installed as close to typical end-use conditions as possible. The test faces consist of a masonry vertical main test face, into which the combustion chamber is located, and a masonry vertical return wall or wing, set at 90° to the main test face. The test specimen should be installed with all the relevant components, and should be assembled in accordance with the manufacturer's instructions. The main test face is at least 8 m high and 2.6 m wide, with the return wing being 8 m high and 1.5 m wide (Figure A2). The distance between the masonry face of the wing wall and the edge of the combustion chamber opening is 250 ± 10 mm with a maximum cladding system thickness of 200 mm. If thicker cladding systems are to be evaluated, the position of the combustion chamber should be adjusted to enable a minimum distance of 50 mm to be maintained between the finished face of the cladding system and the edge of the combustion chamber.

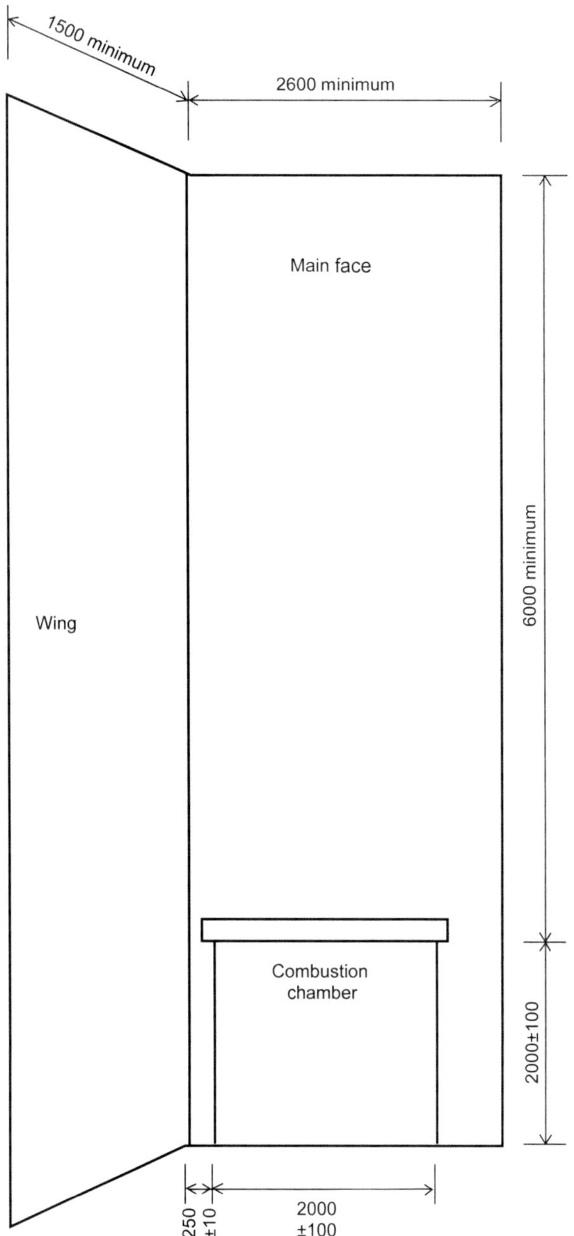

Figure A2: Schematic of test facility (all dimensions are shown in mm)

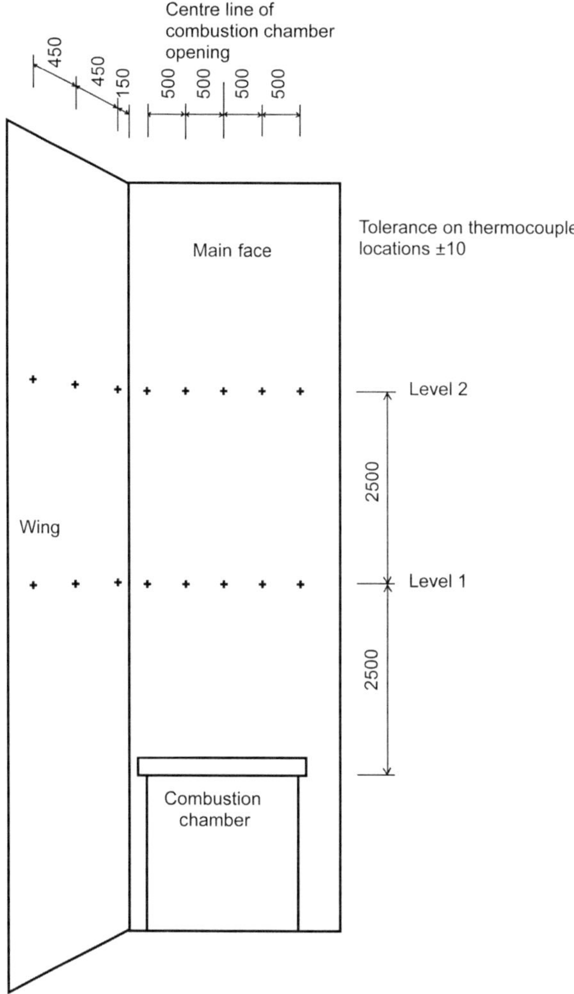

Figure A3: Location of thermocouples (all dimensions are shown in mm)

The test standard provides thermal performance criteria to permit the use of alternative heat sources. A wooden crib is typically used as the heat source for this test, although a gas burner can be used as an alternative. The combustion characteristics of the crib give a total nominal heat output of 4500 MJ over a 30 min period at a peak rate of 3 ± 0.5 MW.

The full details of the test methods can be found in the standard[A1], and although the definitions are repeated above for ease of reference, in all cases the interpretation of test results should be made with regard to the full standard. Figures A2, A3 and A4 show schematics of the test facility and thermocouple locations.

A1.2 Instrumentation

Type K thermocouples are used to monitor temperature at two array locations within the system under test. Figures A3 and A4 summarise the locations of the thermocouples used to monitor the temperatures during the test. At level 2 (Figure A3) the thermocouples are positioned at the mid-depth of each combustible layer, where 'combustible' is defined as referring to a material that does not meet the requirements of Tables A6 or A7 in Approved Document B[A2] – that is, greater than 10 mm thick. Thermocouples are also located at the mid-depth of any cavity that may be present (Figure A4).

A2 PERFORMANCE CRITERIA AND CLASSIFICATION METHOD

The performance criteria and classification method set out in this Annex are based on the latest edition of the BS 8414-1 test method[A1]. In order for a classification to be undertaken, the system must have been tested to the full test-duration requirements of BS 8414-1 without any early termination of the full fire-load exposure period. The primary concerns when setting the performance criteria for these systems are those of fire spread away from the initial fire source, and the rate of fire spread. If fire spread away from the initial fire source occurs, the rate of progress of fire spread or tendency for collapse should not unduly hinder intervention by the emergency services.

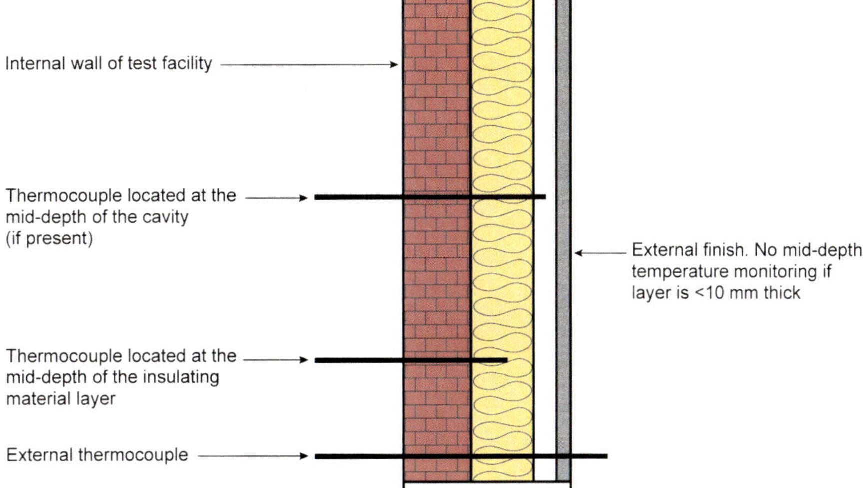

Figure A4: Thermocouple locations within the cladding layers

The performance of the system under investigation is evaluated against three criteria:

- external fire spread
- internal fire spread
- mechanical performance.

The classification applies only to the system as tested and detailed in the classification report. The classification report can only cover the details of the system as tested. It cannot state what is not covered. When specifying or checking a system it is important to check that the classification documents cover the end-use application.

A2.1 Fire-spread start time, t_s

Fire spread is measured by type K thermocouples set at levels 1 and 2 (Figure A3). The start time, t_s, for fire spread is initiated when the temperature first recorded by any external thermocouple at level 1 equals or exceeds a 200 °C temperature rise above the start temperature, T_s, and remains above this value for at least 30 s. An example graph is shown in Figure A5, where ignition of the heat source corresponds to time zero.

A2.2 External fire spread

Failure due to external fire spread is deemed to have occurred if the temperature rise above T_s of any of the external thermocouples at level 2 exceeds 600 °C for a period of at least 30 s, within 15 min of the start time, ts. An example graph is shown in Figure A6.

A2.3 Internal fire spread

Failure due to internal fire spread is deemed to have occurred if the temperature rise above T_s of any of the internal thermocouples at level 2 exceeds 600 °C, for a

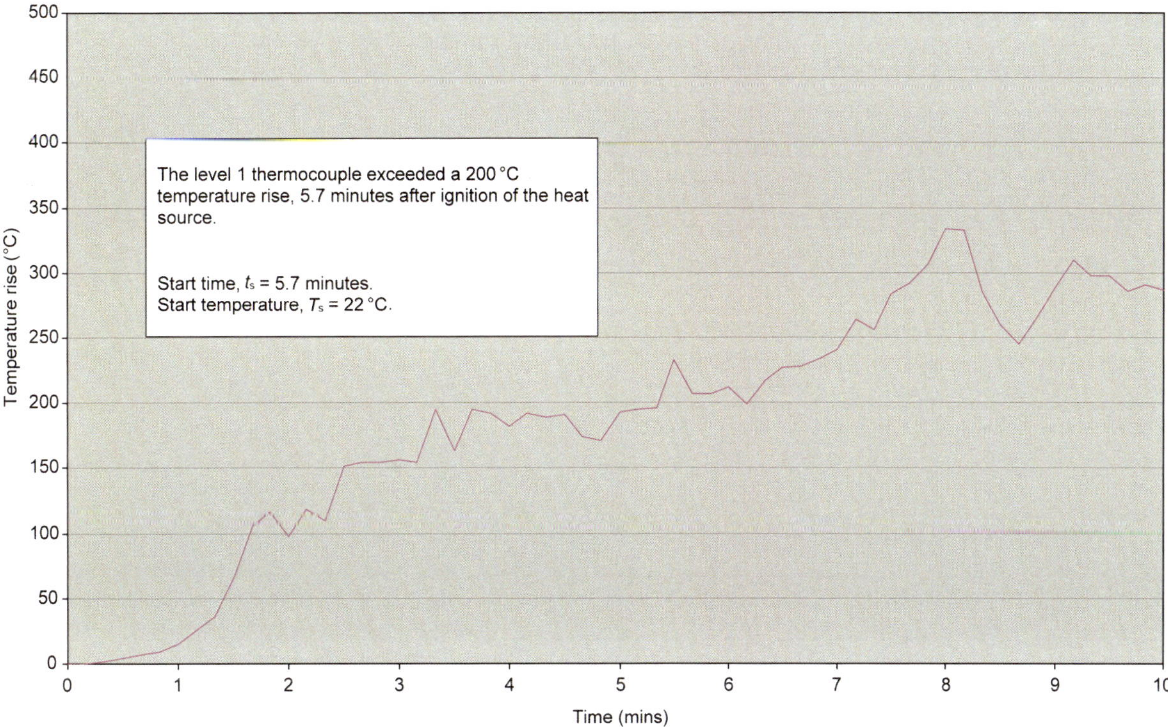

Figure A5: Level 1 thermocouple used to determine start time, t_s

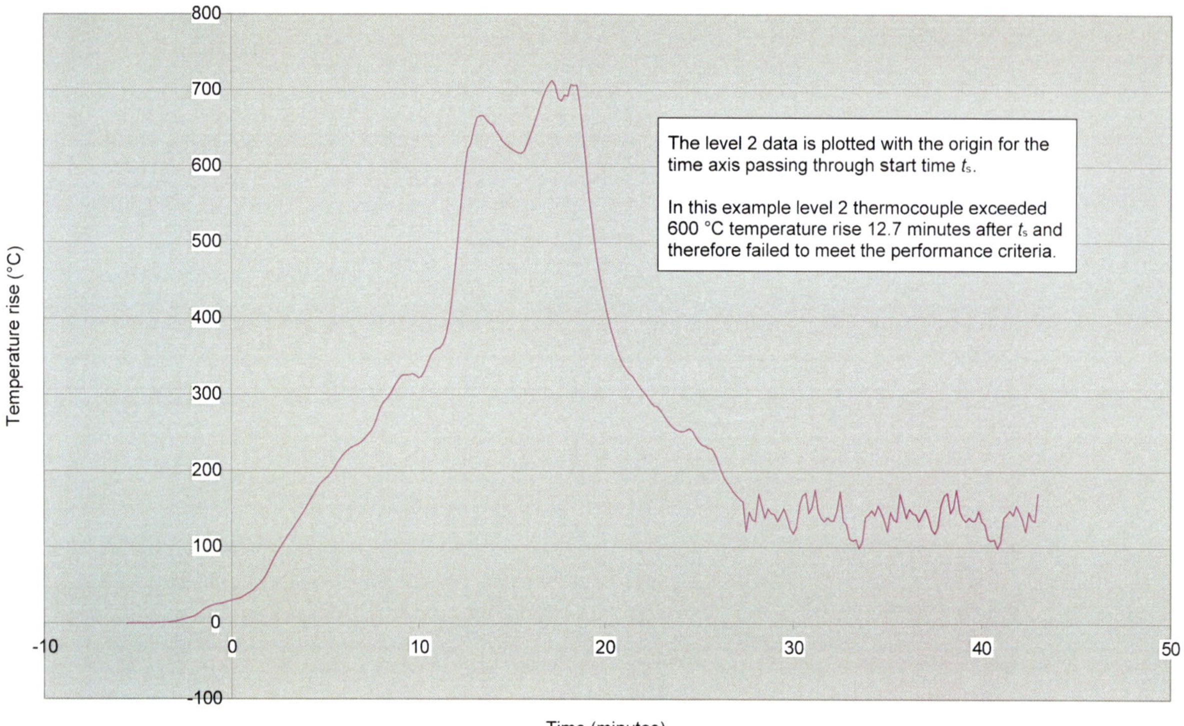

Figure A6: Level 2 thermocouple plotted with start time set to time zero

period of at least 30 s, within 15 min of the start time, t_s. An example graph is shown in Figure A6.

A2.4 Mechanical performance

No failure criteria have been set for mechanical performance. However, ongoing system combustion following extinguishing of the ignition source shall be included in the test and classification reports, together with details of any system collapse, spalling, delamination, flaming debris or pool fires. The nature of the mechanical performance should be considered as part of the overall risk assessment when specifying the system (eg Figure A7).

Figure A7: After the test

A3 REFERENCES

A1 BSI. Fire performance of external cladding systems. Part 1: Test method for non-loadbearing external cladding systems applied to the face the building. BS 8414-1. London, BSI.

A2 Department for Communities and Local Government (DCLG). The Building Regulations (England & Wales) 2010. Approved Document B: Fire safety, Volumes 1 and 2, 2006 edition. London, TSO, 2006. Available from www.planningportal.gov.uk and www.thenbs.com/buildingregs.

ANNEX B: PERFORMANCE CRITERIA AND CLASSIFICATION FOR BS 8414-2

This annex provides a classification system for the test methodology outlined in BS 8414-2 *Fire performance of external cladding systems – Part 2: Test method for non-loadbearing external cladding systems fixed to and supported by a structural steel frame*[B1].

B1 TEST METHOD

Figure B1: A typical test facility

Definitions

Level 1
A height 2 m above the top of the combustion chamber opening (see Figure B3).

Level 2
A height 5 m above the top of the combustion chamber opening (see Figure B3).

Start temperature, T_s
The mean temperature of the thermocouples at level 1 (see Figure B3) during the 5 min before ignition.

Start time, t_s
The time when the temperature recorded by any external thermocouple at level 1 equals or exceeds a 200 °C temperature rise above T_s, and remains above this value for at least 30 s (see Figures B3 and B5).

System
The complete cladding assembly, including any sheeting rails, cavities, fire barriers and weathering membranes or coatings.

B1.1 Principle of test

This test was taken from the scope of BS 8414-2:2005[B1] The test facility has been designed to determine the external fire performance of non-loadbearing external cladding systems, such as glazed elements, infill panels and insulated composite panels, and site-assembled cladding systems, fixed to and supported by a structural steel frame. Figure B1 shows a typical test facility. The principal purpose of the test method is to enable the overall fire performance of the external cladding system, in combination with the relevant substrate wall system and its relevant components, to be assessed as a complete system test as far as is practically possible. The test facility allows external cladding systems to be installed close to typical end-use conditions, and allows variations in the steel frame design to match those used in practice, if required. As in BS 8414-1:2002[B2], the test frame consists of a vertical main test face, in which the combustion chamber is located, and a vertical return wall or wing set at 90° to the main test face. The test specimen is installed to represent typical end-use applications. The test specimen extends a minimum of 6 m above the combustion chamber opening on the main face, and is at least 2.8 m wide. The return wall or wing is the same height as the main test wall, and is a minimum of 1.5 m wide. The distance between the finished face (external surface) of the cladding on the wing and the edge of the combustion chamber opening is 260 (+0 and −100) mm (Figure B2).

The ignition source used in BS 8414-2[B1] is the same as that prescribed in BS 8414-1[B2]: that is, either a wooden crib, as specified in Annex C of BS 8414-1, or a gas burner meeting the characteristics set out in Annex B of BS 8414-1. The full details of the test method can be found in BS 8414-2, and while the definitions are shown for reference, in all cases the interpretation of test results should be made with full regard to BS 8414-2. Figures B2, B3 and B4 show schematics of the test facility and thermocouple locations.

B1.2 Instrumentation

As defined in BS 8414-1[B2] and outlined in Annex A, type K thermocouples are used to monitor temperature at two array locations within the system under test. Figures B3 and B4 indicate the locations of the thermocouples used to monitor the temperatures during the test. At level 2 (Figure B3) the thermocouples are positioned at the mid-depth of any combustible layer present within the system – that is, greater than 10 mm thick – where combustible is defined as a 'layer incorporated in the construction product which, if tested in isolation, would fail the non-combustibility definition in Approved Document B Fire safety'[B3]. Thermocouples are also located at the mid-depth of any cavity that may be present (Figure B4).

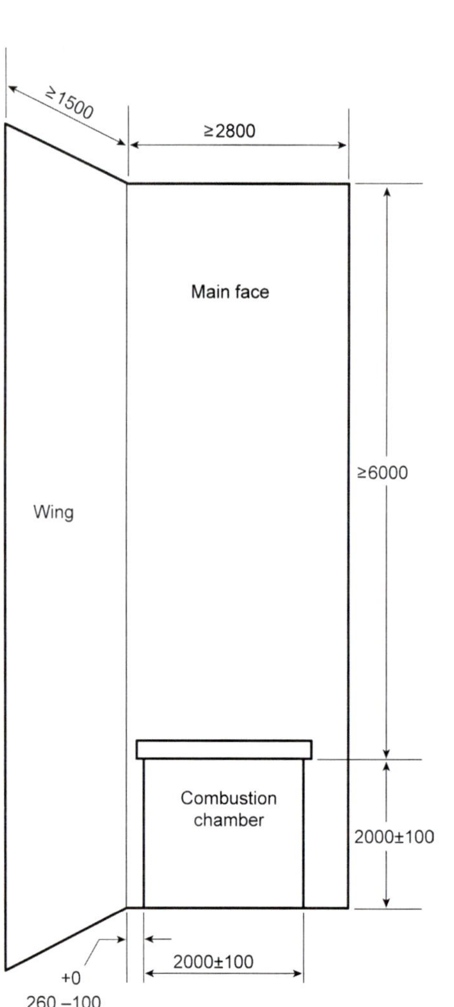

Figure B2: Schematic of typical test facility

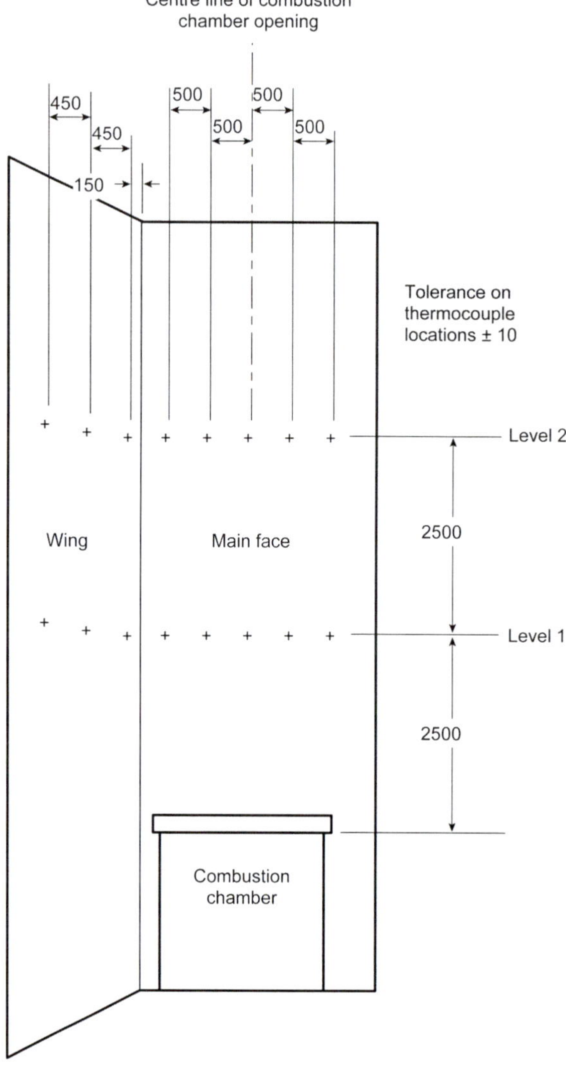

Figure B3: Location of thermocouples (dimensions in mm)

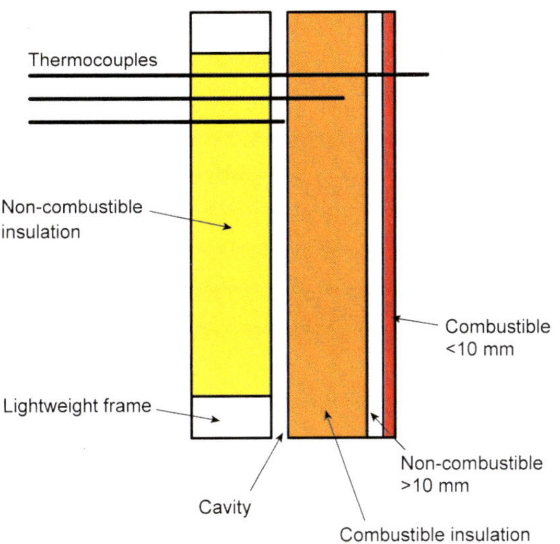

Figure B4: Thermocouple locations through a typical system comprising both combustible and non-combustible materials

B2 PERFORMANCE CRITERIA AND CLASSIFICATION METHOD

The following performance criteria and classification method are based on the latest edition of the BS 8414-2[B1] test method. In order for a classification to be undertaken the system must have been tested to the full test-duration requirements of BS 8414-2 without any early termination of the full fire-load exposure period. As explained in Annex A, the primary concerns when setting the performance criteria for these systems are fire spread away from the initial fire source, and the rate of fire spread. If fire spread away from the initial fire source occurs, the rate of progress of fire spread or tendency for collapse should not unduly hinder intervention by the emergency services. The performance of the system under investigation is evaluated against the following three criteria:

- external fire spread
- internal fire spread
- mechanical performance.

The classification applies only to the system as tested and detailed in the classification report. The classification report can only cover the details of the system as tested. It cannot state what is not covered. When specifying or checking a system it is important to check that the classification documents cover the end-use application.

B2.1 Fire-spread start time, t_s

Fire spread is measured by type K thermocouples set at levels 1 and 2 (Figure B3). The start time for fire spread, t_s, occurs when the temperature first recorded by any external thermocouple at level 1 equals or exceeds a 200 °C temperature rise above the start temperature, *Ts*, and remains above this value for at least 30 s. An example graph is shown in Figure B5, where ignition of the heat source corresponds to time zero (this is also shown as Figure A5 in Annex A, and is repeated here for reference).

B2.2 External fire spread

Failure due to external fire spread is deemed to have occurred if the temperature rise above T_s of any of the external thermocouples at level 2 exceeds 600 °C for a period of at least 30 s, within 15 minutes of the start time, t_s. An example graph is shown in Figure B6 (this is also shown as Figure A6 in Annex A, and is repeated here for reference).

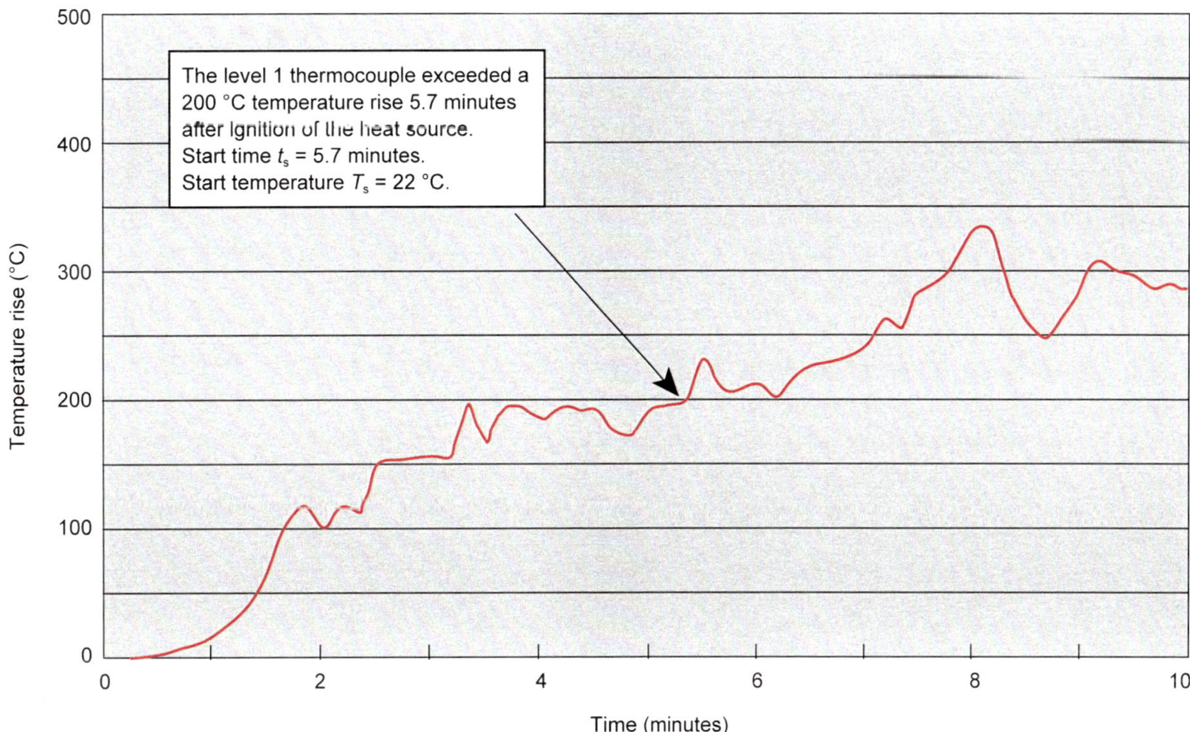

Figure B5: Level 1 thermocouple used to determine start time, t_s

B2.3 Internal fire spread

Failure due to internal fire spread is deemed to have occurred if the temperature rise above T_s of any of the internal thermocouples at level 2 exceeds 600 °C for a period of at least 30 s, within 15 min of the start time, t_s. An example graph is shown in Figure B6 (this is also shown as Figure A6 in Annex A, and is repeated here for reference).

Where system burn-through occurs so that fire reaches the internal surface, failure is deemed to have occurred if continuous flaming, defined as a flame with a duration in excess of 60 s, is observed on the internal surface of the test specimen at or above a height of 0.5 m above the combustion chamber opening within 15 min of the start time, t_s.

B2.4 Mechanical performance

No failure criteria have been set for mechanical performance. However, ongoing system combustion following extinguishing of the ignition source shall be included in the test and classification reports, together with details of any system collapse, spalling, delamination, flaming debris or pool fires. The nature of the mechanical performance should be considered as part of the overall risk assessment when specifying the system.

B3 REFERENCES

B1 BSI. Fire performance of external cladding systems. Part 2: Test method for non-loadbearing external cladding systems fixed to and supported by a structural steel frame. BS 8414-2. London, BSI.

B2 BSI. Fire performance of external cladding systems. Part 1: Test method for non-loadbearing external cladding systems applied to the face the building. BS 8414-1. London, BSI.

B3 Department for Communities and Local Government (DCLG). The Building Regulations (England & Wales) 2010. Approved Document B: Fire safety, Volume 1 Dwelling houses, 2006 edition. London, TSO, 2006. Available from www.planningportal.gov.uk and www.thenbs.com/buildingregs.

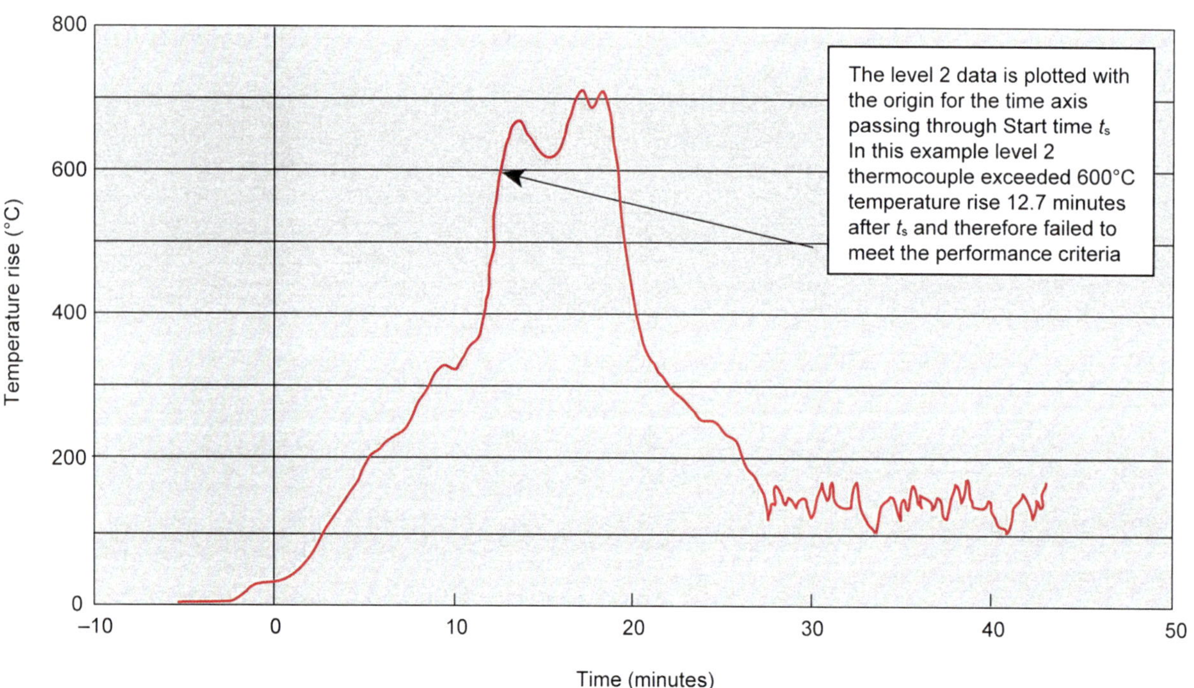

Figure B6: Level 2 thermocouple plotted with start time set to time zero

OTHER TITLES FROM IHS BRE PRESS

Design fires for use in fire safety engineering

Learn how to define a robust and appropriate design fire for use in the fire safety engineering design of a building. This guide presents 29 experimental fires that cover a range of scenarios, and provides data that originate from actual fire tests.

Ref. FB 29

External fire spread

Building separation and boundary distances

Understand the methods for calculating adequate space separation between buildings to ensure that fire spread to an adjacent building is sufficiently delayed for the Fire Service to arrive on site and take preventative action. This new edition of BR 187, in support of national building regulations, gives a range of calculation methods supported by illustrative examples and presents detailed analysis to the methods so that users can create their own fire engineering software.

Ref. BR 187, 2nd edition
Publication due Autumn 2013

Handbook for the structural assessment of large panel system (LPS) dwelling blocks for accidental loading

Get new guidance on the structural assessment and strengthening options for large panel system (LPS) dwelling blocks, focusing primarily on their resistance to accidental loading associated with gas explosions.

For this class of building, this handbook:

- defines the requirements and the criteria against which the results of a structural assessment should be judged
- gives guidance on how to undertake the structural assessments
- explains the risk environment
- details the historic background set in the contemporary philosophical context of the recently introduced structural Eurocodes.

Ref. BR 511

All titles are available in print and pdf format.

Order now @ **www.brebookshop.com** or **phone the IHS Sales Team on +44 (0) 1344 328038**.

OTHER TITLES FROM IHS BRE PRESS

Automatic fire detection and alarm systems

An introductory guide to components and systems

Get an overview of automatic fire detection and alarm systems used in commercial and industrial buildings: how they are designed, how you can manage them and how to ensure that all your regulatory requirements are met.

Ref. BR 510

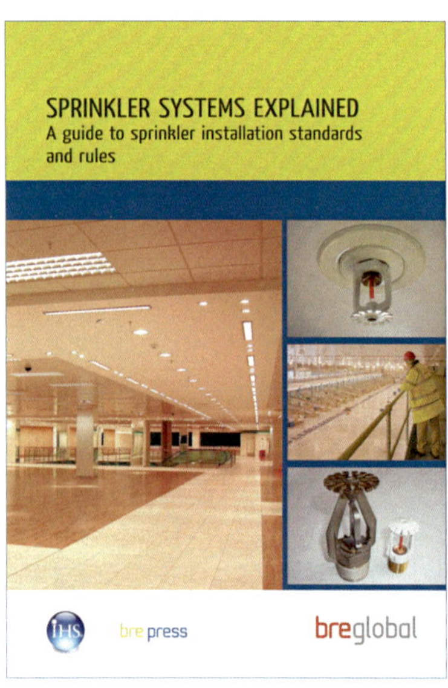

Sprinkler systems explained

A guide to sprinkler installation standards and rules

Understand fire sprinkler installations and the LPC Rules to which they are designed. This guide demystifies the design and use of fire sprinkler installations and explains the engineering behind the rules and regulations and some commone misunderstandings about sprinkler systems.

Ref. BR 503

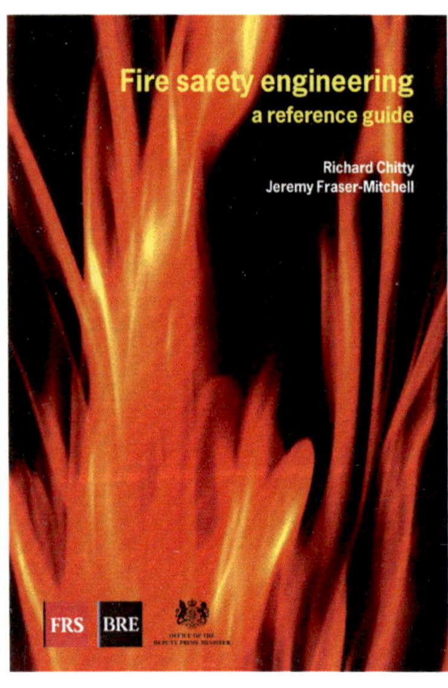

Fire safety engineering

A reference guide

Find out about key topics encountered in fire safety engineering and aspects that should be considered by designers, enforcers and other responsible persons.

Ref. BR 459

All titles are available in print and pdf format.

Order now @ **www.brebookshop.com** or **phone the IHS Sales Team on +44 (0) 1344 328038**.